# Activities

*Help your students engage in inquiry-based learning*

## National Science Education Standards

Several sections in this book cite the applicable standards from the National Research Council's *National Science Education Standards.* Look for the initials NSES as you go through the book. To obtain a copy of *National Science Education Standards,* see **Professional Resources,** page 19.

### General Activity Guidelines

Successful inquiry-based activities, such as those presented below and on the next page, should focus on these abilities:

- Identify questions that can be answered through scientific investigations.
- Design a scientific investigation.
- Develop descriptions, explanations, predictions, and models using evidence.
- Think critically and logically to identify the relationships between evidence and explanations.
- Recognize and analyze alternative explanations and predictions.

## Flowers

### Summary

Students closely examine flowers to identify their parts, focusing on the reproductive organs. This enables them to make the connection between the structure of the flower and the pollination process. From this, students can draw conclusions about how pollen travels from one plant to another as well as the relationship among plants, insects, and birds.

### Materials

**(per student group)**

- hand lens
- flowers—at least six types, three different flowers per pair of students
- drawing paper and colored pencils
- pictures of plant from which flowers are obtained

### NSES

- Reproduction is a characteristic of all living systems; because no organism lives forever, reproduction is essential to the continuation of every species.
- In many species, including humans, females produce eggs and males produce sperm. Plants also reproduce sexually. The egg and sperm are produced in the flowers of flowering plants. An egg and sperm unite to begin development of a new individual. The new individual receives genetic information from the female, its mother (via the egg), and the male, its father (via the sperm). Sexually-produced offspring are never identical to their parents.
- Millions of species of animals, plants, and microorganisms are alive today. Although different species may look dissimilar, the unity among organisms becomes apparent from an analysis of the internal structures; the similarities of their chemical processes; and the evidence of common ancestry.

### Background

About 90 percent of all plants are angiosperms, or flowering plants. The flower is more than a thing of beauty; it is the reproductive organ of the plant whose function is to produce seeds. Much of our food comes from flowering plants. Grasses (corn, wheat, rye, and oats), legumes (pea, bean, peanut), and species of *Brassica* (mustard, cabbage, radish, cauliflower) are all angiosperms. The part of the plant we eat is the result of the flower's successful pollination.

Bisexual flowers have both male and female parts. The stamen is male. This is where the anther produces the pollen. The pistil is female and consists of the stigma, style, and ovary. Unisexual flowers, such as pumpkin flowers, contain a pistil or stamen, but not both. When pollen successfully reaches the pistil, the ovary ripens into a fruit and the ovules inside develop into seeds. Some plants are able to pollinate themselves, but most require pollen from another plant of the same species.

Grasses that have small green or brown flowers are pollinated by wind or water. Large, colorful, and showy flowers attract insects, birds, and animals that carry the pollen from flower to flower. Flowers' varying sizes and shapes attract different types of insects and birds.

### What to Do

1. Ask students to think about flowers, then to draw a detailed picture of a flower.
2. Distribute hand lenses and flowers so that each group has three different ones to examine.
3. Encourage students to observe all parts without damaging the flower. Then have them use colored pencils to draw pictures of each part, showing what was seen with magnification. Remind students to also describe what they saw below each sketch.
4. Invite students to carefully take apart the flower.
5. Ask them to draw and describe all the different parts they observe.
6. Allow a few minutes for students to compare their drawings and observations with other pairs. Did any of them find pollen grains? Explain the function of the stamen, pollen, and pistil.
7. Challenge students to develop an explanation for the differences in petal shape, stamen location, and pistil size. Ask them to think about how the pollen would travel from one plant to another.
8. Encourage students to find pictures of the plants from which the flowers were obtained and to research how the plant is pollinated. ■

## Summary

Students use a hand lens to examine seeds at different stages of early growth. This provides the opportunity to learn more about how plants reproduce and grow, as well as to examine the internal structures that illustrate similarities in species.

## Materials

**(per student group)**

- hand lenses
- 32 petri dishes
- seeds: corn, lima bean, and fast-germinating flowers such as nasturtium

### NSES

- Specialized cells perform specialized functions in multicellular organisms. Groups of specialized cells cooperate to form a tissue, such as a muscle. Different tissues are in turn grouped together to form larger functional units, called organs. Each type of cell, tissue, and organ has a distinct structure and set of funtions that serve the organism as a whole.

## Preparation

1. Label the petri dishes with seed name and hours soaked: 24, 48, 72, and 96 hours.
2. Place 2 to 4 seeds in each petri dish and soak in tap water for the indicated number of hours.
3. Each group of students will get the same seed, soaked for the different time periods.

## Background

Corn is a grass, and belongs to the group of flowering plants called monocotyledons. These have only one seed leaf, or cotyledon, inside the seed. The lima bean and alfalfa are legumes; they belong to the other main group of flowering plants, the dicotyledons. Dicotyledons have two seed leaves inside the seed.

## What to Do

1. Ask students to draw what they think the inside of a seed looks like.
2. Distribute the hand lenses and petri dishes of seeds. Tell students they are to first just examine closely, recording their observations with words and drawings.
3. Invite them to carefully open the seeds so that they can observe the inside. Ask them to again record their observations.
4. Have students move so that each group now can observe another set of seeds. Repeat this until all students have observed the four sets of seeds.
5. Ask students to help complete a Venn diagram comparing and contrasting the seed observed. Challenge students to then develop statements that summarize each seed type.
6. Conclude by distributing the **Reproducible,** page 21, on dicotyledons and monocotyledons. ■

# Bibliography B

*Great books for content, leisure reading, or classroom reference*

## Books for teachers

***American Horticultural Society A–Z Encyclopedia of Garden Plants***
Brickell, Christopher & Judith Zuk. Dorling Kindersley, 1997.

Although designed more for gardeners, this reference boasts more than 15,000 plant descriptions, including thousands of photographs and illustrations.

***A Gardener's Encyclopedia of Wildflowers***
Burrell, Charles Colston. Rodale Press, 1997.

This is a great resource for beginning or advanced gardeners. Use it to identify plant species in the wild as well.

***Introductory Botany: Plants, People, and the Environment***
Berg, Linda R., Ph.D. Harcourt Brace College Publishers, 1997.

A comprehensive guide to plant life, this book also provides information on other aspects of biology such as genetics, earth science, microbiology, and evolution.

***The Plant Book: A Portable Dictionary of the Vascular Plants***
Mabberley, D.J. Cambridge University Press, 1997.

This straightforward dictionary provides definitions for any term that may come up during a unit on plants. ■

## Books for students

***Dictionary of Plant Names: Botanical Names and Their Common Name Equivalents***
Coombes, Alan J. Timber Press, Inc., 1992.

This is a good book for students who wonder how plants get their names. It contains cross-references between common names and scientific names for more than one thousand plants.

***Green Plants***
Tesar, Jenny E. Blackbirch Press, 1993.

As part of the *Our Living World* series, *Green Plants* explains in clear language everything from the history of plants, to their biology, to the role they play in the natural world.

***Janice VanCleave's Plants: Mind-boggling Experiments You Can Turn into Science Fair Projects***
VanCleave, Janice Pratt. Wiley, 1997

This prolific author adds plants to her *Spectacular Science Projects* series. The book's subtitle says it all.

***The Kingfisher Visual Factfinder***
Allaby, Michael, Neil Curtis, Ken Hills, James Muirden & Brian Williams, eds. Kingfisher, 1996.

This resource for students covers history, science, and geography. The sections on plants include interesting text and illustrations.

***Plants***
Alderton, David. Ladybird Books, 1997.

In only 38 pages, this book is packed with concise definitions and explanations; it contains excellent illustrations (including two fold-out sections).

***Plants***
Ganeri, Anita. Franklin Watts, 1992.

Part of the *Nature Detective* series, this book places the emphasis on observation in science while providing background information to help readers form conclusions. ■

# C Careers

*Share this section with students to help them plan for higher education and careers*

## Plant-related Occupations

**Encourage your students to contact experts in plants fields for a classroom visit or to find plant-related question-and-answer sites on the Internet.**

### Botanist

A botanist is a scientist who studies plants. A career in botany requires at least a bachelor's degree for lower-level jobs, such as laboratory assistant. Graduate school—a master's or a Ph.D.—is required for many higher-level botany jobs. A botanist might research plant disease or genetics, find new medicinal uses for plants, or sell plant-based products. A career as a botanist offers many choices: One might work in a laboratory, in the field, or in a corporate office.

To pursue a career as a botanist, it is important to have a well-rounded college preparatory curriculum in high school. Summer or after-school work in a biology-related field would be helpful. Seek out undergraduate botany programs, although you would be wise to pursue a broad curriculum including language arts and social sciences. While you are in college, you will want to conduct a research project, preferably with the guidance of a botanist on faculty. The experience will help you focus on a particular area of interest as well as give you valuable research experience.

Botanical Society of America
1735 Neil Ave., Columbus, OH 43210,
(614) 292-3519, www.botany.org

### Gardener

Professional gardeners do much more than grow plants. Their duties range from designing decorative gardens, to creating food-producing gardens, to protecting plant species. A gardener knows the needs of each plant, such as water, soil, and sunlight. Some of a successful gardener's expertise comes from books, but much of it comes from experience in the garden. Observation is still the most important learning tool for a gardener.

If you want to pursue this career, it is never too early to get experience in the garden. If you don't have a garden at home, seek out a community garden, where volunteers are often welcome. Other opportunities include commercial nurseries or landscaping businesses where full- or part-time work may be available.

American Horticultural Society
7931 East Boulevard Drive, Alexandria, VA 22308
(703) 768-5700, www.ahs.org

### Landscape Architect

Landscape architects work with the community to design parks, public gardens, as well as private and commercial spaces. It is important to have a strong knowledge of plants and a good design sense for this career. Pursue a broad college curriculum, including language arts and social sciences. Many colleges and universities offer undergraduate degrees in landscape architecture. Some people pursue a master's degree in landscape architecture, but there are many programs available. You will be expected to design a research project, a final thesis paper, or both.

American Society of Landscape Architects
636 Eye Street, NW, Washington, DC 20001-3736
(202) 898-2444, www.asla.org

### Florist

Whether you choose to be a grower, a seller, or a designer, you can turn a love of flowers into a successful career as a florist. As a grower, you would need to focus on flowers' optimal growing conditions. As a retailer, you would need to stay abreast of the latest trends and learn which flowers are best suited for specific situations and environments. Designers use artistic flair to create floral arrangements that add beauty to any event.

While formal education is not necessary for a career as a florist, it is certainly valuable. Other experiences worth pursuing as an aspiring florist are internships with local growers, retailers, or floral designers.

Society of American Florists
1601 Duke Street, Alexandria, VA 22314
(800) 336-4743, www.safnow.org

# Demonstrations

*The excitement's up front with you, the teacher, for these classroom "events"*



## General Demonstration Guidelines

For safety or materials reasons, demonstrations need teacher direction. But they can still provide students with opportunities to develop abilities and understandings of inquiry, especially when the demonstration replicates a "great discovery."
Questioning strategies should focus on developing the following abilities of inquiry:
- Design a scientific investigation.
- Develop descriptions, explanations, predictions, and models using evidence.
- Think critically and logically to identify relationships between evidence and explanations.
- Recognize and analyze alternative explanations and predictions.

## Fruits and Vegetables

### Summary

Students examine different fruits to discover the pattern of seeds within. This provides an opportunity to learn more about how plants reproduce. Students should also observe similarities in the internal structures that illustrate similarities among plant families.

### Materials

- fruits including banana, kiwi, tomato, cucumber, apple, orange, avocado, squash, green beans
- other plant foods including lettuce, cauliflower, celery, green onion, potato, carrot
- sharp knife to cut the plant foods

### NSES

- Reproduction is a characteristic of all living systems; because no organism lives forever, reproduction is essential to the continuation of every species.
- In many species, including humans, females produce eggs and males produce sperm. Plants also reproduce sexually. The egg and sperm are produced in the flowers of flowering plants. An egg and sperm unite to begin development of a new individual. The new individual receives genetic information from its mother (via the egg) and the father (via the sperm). Sexually produced offspring are never identical to their parents.
- Millions of species of animals, plants, and microorganisms are alive today. Although different species may look dissimilar, the unity among organisms becomes apparent from an analysis of the internal structures; the similarities of their chemical processes; and the evidence of common ancestry.

### What to Do

1. Ask students to list as many fruits as possible. Then ask them to describe what characteristics led them to conclude these items were fruits.
2. Show your collection of fruits. Have students describe the outside of each. How are they similar? How are they different?
3. Show the other foods and again have students describe similarities and differences.
4. Cut each fruit into slices that allow the seed pattern to be easily observed. Place slices of different fruits on paper plates. Repeat this with the other foods.
5. Give each group a plate of fruits and a plate of other foods. Challenge students to find similarities and differences within each group and between the two groups.
6. Ask students to record their observations with words and drawings, and then create a Venn diagram to summarize their findings.
7. Conclude by discussing the other foods on their lists of fruits. ■

# E Experiments

*To provide students with guided practice in inquiry*

Discovery Channel School

## General Experiment Guidelines

The teacher should provide the skeleton of the experiment, but students still should be involved in making decisions regarding data collection, controlling variables, and depth of study. The aim of the experiments should be to focus students on these abilities:

- Design (or modify) a scientific investigation.
- Use appropriate tools and techniques to gather, analyze, and interpret data.
- Develop descriptions, explanations, predictions, and models using evidence.
- Think critically and logically to identify the relationships between evidence and explanations.
- Recognize and analyze alternative explanations and predictions.
- Communicate procedures and explanations.
- Use mathematics.

## Plant Care

### Materials

**(per student group)**

- film canisters or planting trays
- hand lens
- centimeter ruler
- journals
- graph and drawing paper

### NSES

- Living systems at all levels of organization demonstrate the complementary nature of structure and function. Important levels of organization include cells, organs, tissues, organ systems, whole organisms, and ecosystems.
- Reproduction is a characteristic of all living systems; because no organism lives forever, reproduction is essential to the continuation of every species.
- In many species, including humans, females produce eggs and males produce sperm. Plants also reproduce sexually. The egg and sperm are produced in the flowers of flowering plants. An egg and sperm unite to begin development of a new individual. The new individual receives genetic information from its mother (via the egg) and the father (via the sperm). Sexually produced offspring are never identical to their parents.

### Background

Grasses provide food for much of the world. Rice, wheat, corn, and rye are all grasses.

### What to Do

1. Ask students to think about what plants need to grow and reproduce.
2. After allowing them to think for a few minutes, invite students to share their ideas.
3. Engage students in a discussion of the variables that affect a plant's growth and development. These could include soil type, amount of water, amount of sunlight, sunlight versus artificial light, and crowding. Ask them to think of ways these variables could be tested.
4. Review with the class how to set up an experiment that controls for one variable. Ask each pair of students to identify a variable to test and how they will collect and record data. Suggest that root growth is an important factor.
5. Work with students to plan plant care over the weekends.
6. Remind students to record quantitative and qualitative data daily. Routinely engage students in a discussion of their findings.
7. When the time span for the experiment is over, have students develop conclusions that address how the variable they tested affects the plant's growth.

**Note:** The Wisconsin Fast Plant™ kits allow students to perform seed tests that include reproduction or yield. (See **Professional Resources,** page 19.) Information about Fast Plants and instructional materials using them are available under the Fast Plants Learning Ecosystem Web site. Log on to www.fastplants.org ■

# Aquatic Plants

## Materials

**(per student group)**

- duckweed
- hand lens
- water—pond, tap, distilled
- clear plastic cups or petri dishes
- journals, graph paper, drawing paper, and colored pencils

## Background

Duckweed and eel grass are two common aquatic plants that depend on water to pollinate seeds for reproduction. They also are indicators of water quality.

## What to Do

1. Ask students to think about plants that grow in water. Do they have the same structures as land plants? Where do they live? What do they use for food? How big do they get? How do they reproduce?
2. After allowing students time to think and record their ideas, distribute hand lenses and duckweed placed in clear cups to each pair of students. Each cup's label should indicate the type of water. Have students record detailed observations.
3. Then create a list of questions they could use in designing an experiment that uses duckweed: How big does it grow? What kind of water results in the most growth? How rapidly does the plant reproduce? How big are its flowers?
4. Identify the variables students can control: heat, light, space, type of water.
5. Allow students time to plan their investigation, including data tables and method of collecting data.
6. Allow at least a week for the data collection. Then engage students in a discussion of their findings. Did any groups see a flower? How was the flower pollinated?
7. Conclude by having them write up their experiments, incorporating research about aquatic plants. ■

**NSES**

- Reproduction is a characteristic of all living systems; because no organism lives forever, reproduction is essential to the continuation of every species.

F

# Free Stuff

*Charge up your Plants unit—at no extra charge!*

## Pictures

The Botanical Society of America has an extensive online gallery of images available on their Web site. Approximately 800 images in the gallery are divided into categories such as Plant Geography, Carnivorous Plants, Lichens, and Plant Anatomy. The images are copyrighted, but they may be downloaded and used for educational purposes. Go to http://images.botany.org ■

## Wondrous Web Sites

The Internet offers a plethora of free, downloadable lesson plans on almost any topic. The following URLs contain quality lesson plans that can be used to enhance your Plants unit.

Engaging Science includes a lesson-plan exchange. The Life Science section provides several good lesson plans that incorporate the study of plants. You can find these lesson plans by going to www.engagingscience.org/lpe/lpe.html and clicking on Life Science.

A middle school science teacher offers a collection of lesson plans at www.geocities.com/damselflywings

Two science television shows for kids have good Web sites where you'll find useful lesson plans. The PBS science program *Newton's Apple* makes its lesson plans available for educational use; several are related to plants. Of particular interest is a lesson on photosynthesis, located at www.eecs.umich.edu/mathscience/funexperiments/agesubject/lessons/newton/phytosy.html

The *Beakman's World* site offers answers to "50 Terrific Questions" with fun lesson plans. Find out what cutting down trees has to do with global warming at www.beakman.com/warm/warm.html ■

## Note Cards

The Nature Conservancy offers a set of beautiful note cards showcasing plant and animal species the group works to protect. Visit the Fun and Free area of their Web site at www.tnc.org ■

**www.discoveryschool.com**

The URLs listed here can be accessed through www.discoveryschool.com for your convenience. Go to **Science Collections** and follow the links.

## All About Plants Crossword

**See p. 32 for answers.**

**Across**

1. Substance found in plants that makes them green
2. Kind of plant, such as pine or fir, that produces cones
3. Only plant with jaws, although not the only plant to eat insects
4. These are onecelled and, like plants, they also photosynthesize.
5. Tree that loses its leaves at the end of the season
6. Process by which a plant makes its own food using the air, water, and sunlight
7. Plants use carbon __________ from the air to make food.
8. Most of a plant's food is made in this part.

**Down**

1. This slow-growing alga-fungus is common in the coldest areas on Earth.
2. Usually underground, plant part that absorbs water and nutrients from the soil
3. Where flowers produce pollen
4. Where pollination takes place in a flower
5. Outside part of a leaf; its skin
6. Tubes that carry water from a plant's roots to its leaves

## Poison Plants Mix-up

**Unscramble the names of these commonly found poisonous plants.**

1. A shrub that has a variety of bright-colored blooms. LAAZEA
2. This yellow-and-white flower blooms in the spring. LIFOFDAD
3. A deadly plant that can be used to treat heart disease. Its name sounds like an animal's accessory. VLOFEXOG
4. A holiday plant associated with good cheer, but if eaten, it can cause severe stomachache. EITLETOMS
5. This holiday plant bears red berries and prickly leaves. LYHLO
6. Touch this plant and your skin may itch. NOSPIO YVI

# Homework Helpers

*Build a variety of assignments around this information*

Discovery Channel School

## Plants Information You Can Use

### Photosynthesis

Plants make their own food with photosynthesis. During this process a plant uses water, carbon dioxide from the air, and energy from sunlight to make food. Plants and algae are the only living things that can make their own food in this way. In the process of making food, plants provide a valuable resource for the earth. Plants take in carbon dioxide, which can be dangerous to humans and other living things if it builds up in the air. Also, they release oxygen into the air, which allows other living things to breathe.

Plants are perfectly designed for this purpose. Think about the ways in which a plant seems specially designed to collect water. What effect do you think the shape of a leaf has on the process of photosynthesis and on the overall survival of the plant? Oxygen is a product of photosynthesis. What happens during the process of photosynthesis to release oxygen? How does this happen? ■

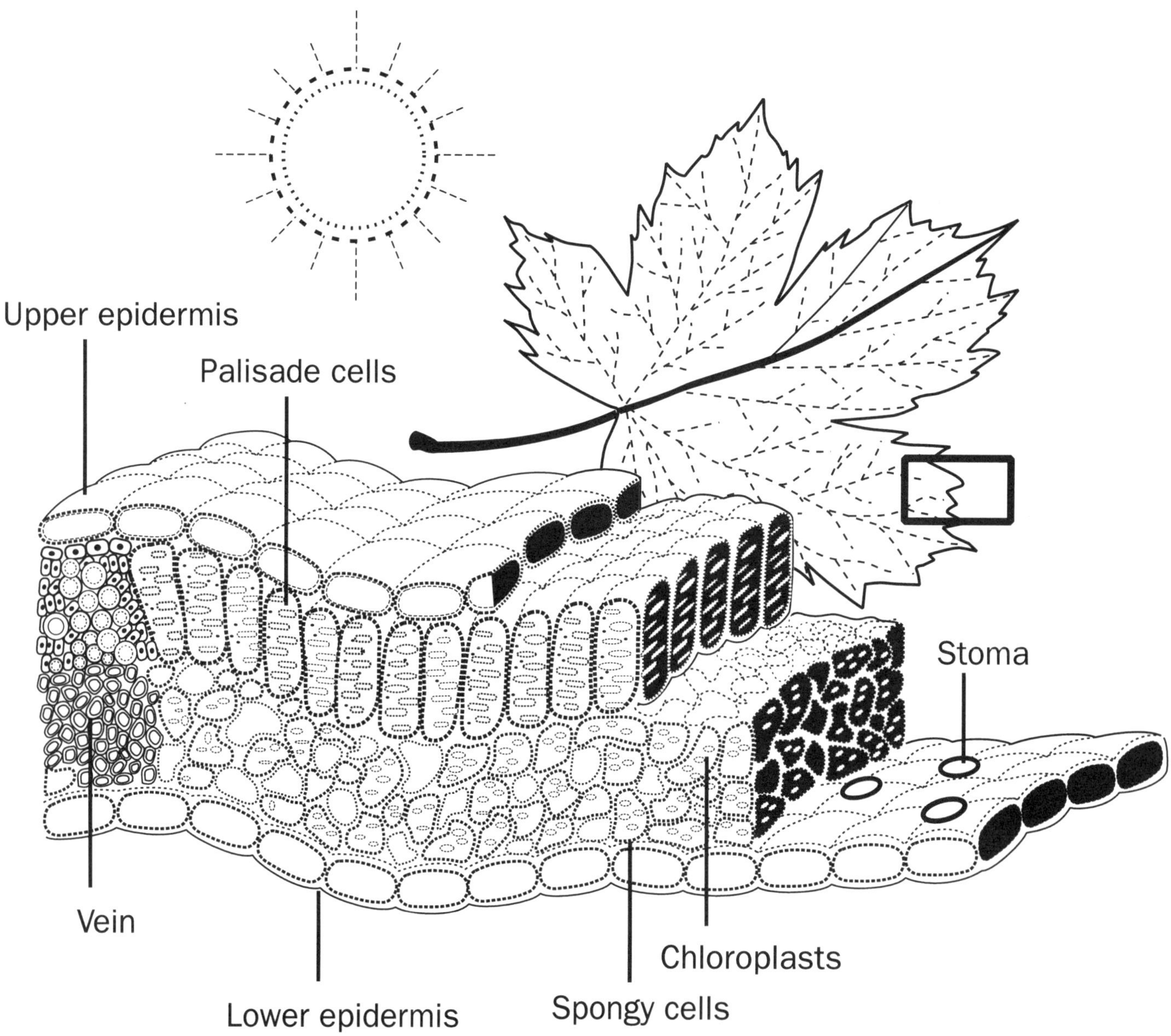

# Interdisciplinary

*Connections to other subjects add new dimensions to science*



**National Science Education Standards**

- The program of study in science for all students should be developmentally appropriate, interesting, and relevant; emphasize student understanding through inquiry; and be connected with other school subjects.

## Social Studies: Plants in Your Community

We already know the positive effects plants have on the earth: cleaning the air we breathe and providing food. But plants also play a social role in communities. Public parks, arboretums, and community gardens often bring people together for events. Forested areas provide peaceful places for hikes and nature walks. Have your students brainstorm and research the roles plants play in their own community. Students can survey their hometowns and interview local residents and officials about their thoughts on public parks and other green areas. After students conduct research, invite them to create a report for the town or city. They may also launch a campaign to persuade local officials to set aside more land for public parks or gardens. ■

## Language Arts: Plant Life as Inspiration

Throughout history, nature has inspired great poets and writers. Henry David Thoreau wrote about his two years, two weeks, and two days spent observing nature in Concord, Massachusetts. Emily Dickinson, Walt Whitman, and Robert Frost are writers who described nature and the natural world. Invite your students to make a list of other writers and poets, either modern or historical, who were inspired by nature. Encourage students to collect examples of the writings of each of these authors for display on a "Nature in Literature" bulletin board. ■

# Interdisciplinary

*Connections to other subjects add new dimensions to science*

Discovery Channel School

## National Science Education Standards

- The program of study in science for all students should be developmentally appropriate, interesting, and relevant; emphasize student understanding through inquiry; and be connected with other school subjects.

## Math: How Does Your Garden Grow?

Planting a garden can be very costly. Your students can participate in this exercise to learn about budgeting and problem solving. Set a spending limit for each garden, say two hundred dollars. Have your students plan their individual gardens, keeping the budget in mind. Students should sketch their plans first. Encourage students to consider design, as well as personal likes and dislikes. Students can then draw a more concrete plan for their gardens. Students should then shop online or visit a local nursery to determine the cost of the plants they have selected. At this point, students can begin to work with their plan to keep the garden within the budget while retaining their creative vision. ■

## History

When early settlers came to North America from Europe, they discovered that native plants were different from those in their homelands. They may have shared plants with Native Americans, and they certainly introduced new species of wheat and rye. Invite students to research plants that were introduced to North America. Then ask them the following questions: What are the differences between native, wild, and domesticated plants? What were the benefits or pitfalls of introducing new plant species to North America (or elsewhere)? How did these plants adapt to a new environment? Do these plants help or harm the environment? Explain why. What are the benefits and costs of plant diversification? ■

# Just for Fun ☺



*A little lightheartedness leavens any learning!*

## Make Your Own Ink

**Save these for a rainy day—or anytime.**

Throughout history, people have figured out ways to use plants. In the past when people wanted to write, they would dip a fountain pen into a bottle of ink, which was often made from dark berries. Such ink is easy to make. Different berries make different ink colors, so be sure to have enough jars or other containers on hand. Protect surface areas, as berries can stain clothing. Place the berries in a plastic cup and mash them with a spoon. Place a paper towel or cheesecloth over the opening of a jar. Drain the liquid from the berries into the jar. When all the liquid has drained, discard the berries. The remaining liquid is the ink. Note: Buy an inexpensive fountain pen, available at most stationery stores, to use with your ink. (This activity could damage it somewhat.) Follow the directions below to make your own paper, and you'll be ready to write using plant products. ■

## Make Your Own Paper

Most of the paper we use comes from trees, but more and more paper is made from recycled materials. Here's a way you can make paper at home using tissues. You'll need a piece of screen measuring about 8 × 12 inches. Fold down the edges of the screen to form a sturdy frame and to blunt any sharp edges. Shred about five tissues and place them in a mixing bowl. Pour enough hot water into the bowl to generously cover the tissues. Using an electric mixer, beat the tissue-and-water mixture; add a tablespoon of starch. (You can find this at a grocery store.) Place the screen you made into the bottom of a baking pan. Pour the tissue-and-water mixture over the screen into the pan. Swish the screen around to get an even layer of tissue over it, and lift the screen out of the pan, allowing the excess water to drain off of it. Place the screen on a small stack of newspapers with the tissue side up. Cover the screen with a thin layer of newspaper and use a rolling pin to squeeze out the remaining water. When most of the water has been squeezed out, lift off the top newspaper. Carefully peel your newly made paper off the screen and place it on the newspaper to dry. When the paper is dry, use your homemade berry ink to write on it. ■

## Jokes, Riddles, and Puns

Knock, knock
Who's there?
Conifer
Conifer who?
*I conifer got what the question was.*

What's a tree's favorite drink?
*Root beer*

What did one dandelion say to the other?
*Take me to your weeder.*

What kind of flower grows on your face?
*Tulips*

Why did the tree go to the dentist?
*Because it needed a root canal.*

How do you send a message in the forest?
*By moss code*

What kind of tree is always jealous?
*An evergreen*

# K Key Concepts

*Content standards to be addressed in teaching Plants*

Discovery Channel School

## NSES for Plants

**These concepts are adapted from content standards for grades 5–8, *National Science Education Standards***

### Structure and Function
(pp. 156–157)

- Living systems at all levels of organization demonstrate the complementary nature of structure and function. Important levels of organization include cells, organs, tissues, organ systems, whole organisms, and ecosystems.
- All organisms are composed of cells—the fundamental unit of life. Most organisms are single cells; other organisms, including humans, are multi-cellular.

### Reproduction and Heredity
(p. 157)

- Reproduction is a characteristic of all living systems; because no organism lives forever, reproduction is essential to the continuation of every species.
- In many species, including humans, females produce eggs and males produce sperm. Plants also reproduce sexually—the egg and the sperm are produced in the flowers of flowering plants. An egg and sperm unite to begin development of a new individual. The new individual receives genetic information from its mother (via the egg) and the father (via the sperm). Sexually produced offspring are never identical to their parents.
- Every organism requires a set of instruction for specifying its traits. Heredity is the passage of these instructions from one generation to another.

### Diversity and Adaptations of Organisms (p. 158)

- Millions of species of animals, plants, and microorganisms are alive today. Although different species might look dissimilar, the unity among organisms becomes apparent from an analysis of the internal structures, the similarities of their chemical processes, and the evidence of common ancestry.
- Extinction of a species occurs when the environment changes and the adaptive characteristics of a species are insufficient to allow its survival. Fossils indicate that many organisms that lived long ago are extinct. Extinction of a species is common; most species that have lived on the earth no longer exist.

### Populations and Ecosystems
(pp. 157–158)

- Populations of organisms can be categorized by the function they serve in an ecosystem. Plants and some microorganisms are producers—they make their own food. All animals, including humans, are consumers, which obtain food by eating other organisms. Decomposers, primarily bacteria and fungi, are consumers that use waste materials and dead organisms for food. Food webs identify the relationships among producers, consumers, and decomposers in an ecosystem.
- For ecosystems, the major source of energy is sunlight. Energy entering ecosystems as sunlight is transferred by producers into chemical energy through photosynthesis. That energy then passes from organism to organism in food webs. ■

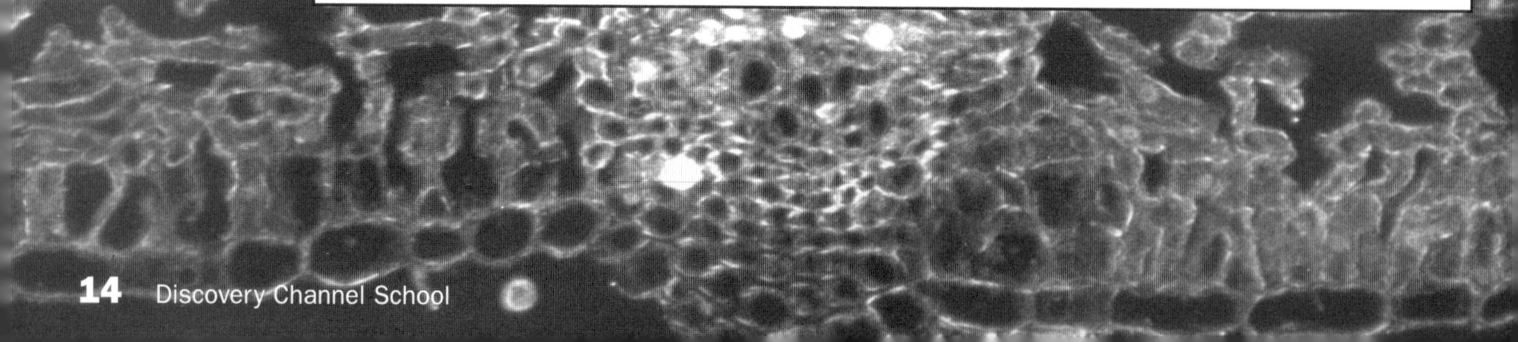

# Leisure Activities

*Reinforce the study of Plants in your students' "off" hours*



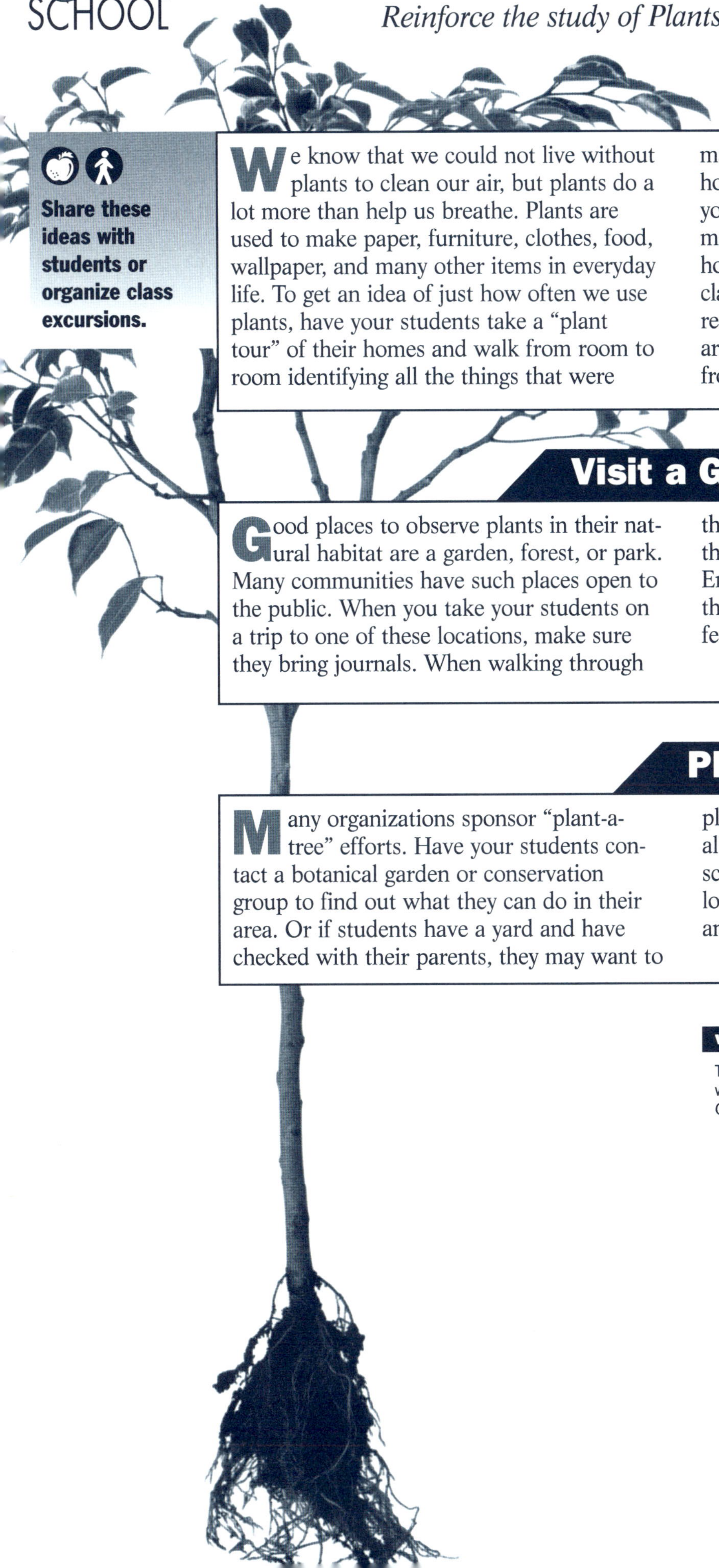

**Share these ideas with students or organize class excursions.**

## Plants All Around

We know that we could not live without plants to clean our air, but plants do a lot more than help us breathe. Plants are used to make paper, furniture, clothes, food, wallpaper, and many other items in everyday life. To get an idea of just how often we use plants, have your students take a "plant tour" of their homes and walk from room to room identifying all the things that were made from plants. To prepare for this at-home activity, take students on a plant tour of your school or classroom. Have students make a list of everything that they find in their homes. Students can share their lists with the class. As an extra exercise, students may research and list clothing and other items that are now made artificially, but were once made from plant materials. ■

## Visit a Garden, Forest, or Park

Good places to observe plants in their natural habitat are a garden, forest, or park. Many communities have such places open to the public. When you take your students on a trip to one of these locations, make sure they bring journals. When walking through the garden or park, students should record their observations of plants in their journals. Encourage students to note the diversity of the plants and any interesting structural differences among the plants. ■

## Plant Trees in Your Area

Many organizations sponsor "plant-a-tree" efforts. Have your students contact a botanical garden or conservation group to find out what they can do in their area. Or if students have a yard and have checked with their parents, they may want to plant a tree in their own yard. You could also organize a tree-planting day at your school. Suggest that students approach a local nursery to see if the owners will donate any trees to the school. ■

**www.discoveryschool.com**

The URLs listed here can be accessed through www.discoveryschool.com for your convenience. Go to **Science Collections** and follow the links.

# Misconceptions

*Common—and mistaken—ideas about Plants*

Discovery Channel School

## Don't You Believe It!

**Set the record straight with your students.**

### A tomato is a vegetable.

When you think of fruit, what do you think of? Probably something sweet and juicy, such as an orange, apple, or strawberry. You probably don't think of a tomato. Do you have tomato pie a la mode for dessert? Of course not. Because most fruits are sweet or tart, the tomato doesn't seem to fall into that category. But it is a fruit. The fruit of a plant is designed to protect the seeds of that plant and help in seed dispersal. The fruit becomes ripe and opens up. From there the seeds may be dispersed by the wind or by animals. Other fruits often thought of as vegetables are cucumber, pepper, zucchini, and eggplant. In general fruits are sweet-tasting. Ask students to consider why a sweet-tasting fruit would aid in seed dispersal.

### Trees have red, orange, and yellow leaves only in the fall.

You may think that this is true because you've observed this change. Although the leaves "change" in the fall, chlorophyll hides their brilliant colors.

Chlorophyll makes leaves green. It is responsible for photosynthesis, the process by which a plant makes its own food. When winter approaches, deciduous trees prepare for the winter. The trees conserve food and water for the harsh weather ahead, so the tree stops sending food to the leaves. Without a food source, the leaves begin to lose their chlorophyll. As the chlorophyll is gradually reduced in a leaf, its other colors begin to show. So leaves do not get those bright colors only in the fall. They simply lose their green! ■

# Noteworthy People

*Share these achievers' stories with your students*

N

**Students might like to research projects or presentations on other achievers including**

- **Jean-Baptiste Lamarck (French botanist)**
- **Charles Darwin (natural selection)**
- **Nathaniel Lord Britton (Botanical Society of America)**

**American botanists**

- **Katherine Esau**
- **Judith Croxdale**
- **Paul Williams**

**National Science Education Standards**

- Many individuals have contributed to the traditions of science. Studying some of these individuals provides further understanding of scientific inquiry, science as a human endeavor, the nature of science, and the relationships between science and society.

## Carolus Linnaeus (Carl von Linné)

### May 23, 1707–Jan. 10, 1778

Linnaeus was born in Sweden and developed a love of plants at a young age. He studied plants wherever he went. As a botany student at the University of Uppsala in Sweden, he discovered flaws in the system used to organize plants. In 1751 he published *Philisophica Botanica,* which described his new classification system. He also published *Species Plantarum* (1753) and *Systema Naturae.* Linnaeus based his system on the physical characteristics of plants, such as stamens and pistils. He also developed a new, simplified plant-naming system. Botanists still use that system today.

Like many great scientists, Linnaeus built on the work of other scientists. In the mid-1500s, scientists began to explore and observe their subjects. This was the Renaissance, a period of intense exploration and discovery. During this period, the scientist Gaspard Bauhin developed the concept of genus, or the idea that groups of species share similar characteristics and may come from a common ancestor. This concept was controversial at the time, but it laid the groundwork for Linnaeus's work 200 years later.

Linnaeus's system made it possible for scientists to keep track of all the known species, a task that would otherwise be impossible. With such great diversity of plants, classification continues to be a vital part of botany. In 1761 the Swedish royal family granted Linnaeus nobility for his work and he became Carl von Linné. ■

## Gregor Mendel

### July 22, 1822–Jan. 6, 1884

Gregor Mendel introduced the concept of heredity. He experimented with pea plants, discovering that specific characteristics were passed down through generations. His conclusions formed the basic laws of heredity. Before Mendel's work, scientists believed that a plant's environment determined its traits.

Mendel was born in Moravia, known today as the Czech Republic, in Eastern Europe. He was an excellent student as a young man and had an innate interest in nature, but his family could not afford to send him to the university. Instead he entered an Augustinian monastery to continue his studies. As a monk, Mendel began teaching natural science to high school students. He would often stroll the monastery gardens, observing plants and forming questions about how plants obtained their characteristics. It was during one of these strolls that he noticed an atypical plant variety. His curiosity led him to conduct an experiment: He planted the atypical plant next to a typical plant to see if their offspring would pick up any characteristics of a neighboring plant. He found that a plant's offspring retained its parents' characteristics. He theorized that a plant's proximity to other plants had no effect on its growth or appearance. This went against many of the accepted theories of the time.

Mendel continued his experiments with pea plants, deriving the fundamental ideas about heredity. He found that rather than blending, hereditary factors are passed distinctly from one generation to the next. He also discovered that each parent passes only half of its traits to its offspring, and offspring will receive different traits from each parent. In addition, he posited the existence of dominant (versus recessive) traits. Although unappreciated at the time, Mendel's work formed the foundation for all subsequent work in genetics. ■

## Cool Plant Facts

### Study plants by observation: How does their appearance matter?

Take a walk through a garden or park and really look at the plants. Are certain plants brightly colored and others green? Do some plants smell good, others bad, while some seem to have no fragrance at all? Do some grow close to the ground, while others grow tall? Even without formally studying plants, you can figure out that plants have certain characteristics for very specific reasons. Just using your observations, try to determine how each characteristic might contribute to a plant's survival.

### Thanks but no thanks: Exotic species can be an unwelcome addition.

"Don't mess with Mother Nature." People have ignored that advice throughout history, sometimes with disastrous outcomes. In every area of the world there are indigenous, or native plants. For a variety of reasons, people have introduced new plants into an environment. When plants are grown in a new area, the results may be uncontrollable.

European settlers arrived in America, carrying plants with them. They brought the garlic mustard plant from Europe for its high vitamin content and good flavor. They also planted it to prevent soil erosion. But these early gardeners did not realize that garlic mustard has no natural enemies in this new land, so it spread very quickly. Today it grows out of control and chokes other plants by denying them sunlight, water, and the nutrients necessary for survival.

In the Midwest, where garlic mustard is especially prolific, scientists are working on ways to control the plant. One idea is to import weevils that feed on it exclusively. However, the federal government must approve such a plan, because it is also risky to introduce a new animal species. Despite this example, the introduction of non-native domesticated plants usually has been successful.

### Phytoremediation: Using Plants to Clean up Hazardous Materials

Hazardous materials in the ground are a big problem all over the world. These potentially dangerous substances—explosives, chemicals, and oils—can seep into groundwater and harm people or animals. Often a problem is fixed by digging up the materials and burying them in a less harmful place. But this is really an expensive Band-Aid™.

Scientists are beginning to rethink a process that has been around for a long time: phytoremediation, whereby plants destroy or absorb contaminants from the soil. Some forms of phytoremediation employ natural processes to extract minerals from the soil. At Chernobyl, Russia, the site of the worst nuclear power-plant accident in history, sunflowers absorbed radioactive material from standing water. In other cases of phytoremediation, the roots of a plant attract microorganisms that eat the contaminants. Around oil wells in Arkansas, the ground is saturated with oil. Scientists plant crabgrass to clean up the oil. Crabgrass roots attract bugs, which consume the oil.

Phytoremediation is a slow process. If people are in imminent danger, a faster form of remediation may be more appropriate. However in many cases, phytoremediation acts as an important means of cleaning up toxic sites. ■

# Professional Resources

*Peer groups, programs, associations, and more*



## National Science Education Standards

### National Research Council

National Academy Press

The National Science Education Standards book, developed by thousands of teachers, scientists, science educators, and other experts, offers a comprehensive guide to teaching science in schools today. The standards and teaching practices in this resource provide a coherent vision of what it means to be scientifically literate, describing what all students in grades K–12 should understand and be able to do in various science categories. To order contact

National Academy Press
2101 Constitution, NW, Lockbox 285
Washington, DC 20005
(800) 624–6242, (202) 334-3313, www.nas.edu ■

## Educational Resources

### National Science Teachers Association

1840 Wilson, Arlington, VA 22201-3000
www.ntsa.org, (703) 243-7100

The NTSA publishes newsletters, five journals, many books, a new children's magazine called *Dragonfly,* and many other publications. NSTA conducts national and regional conventions in the spring and fall. Their Web site provides a position statement on the teaching of evolution, and links to many additional resources.

Also be sure to check out the Science Teacher Association in your state. State conferences provide quality workshops and teaching materials, as well as exhibits from publishers and local organizations.

### Wisconsin Fast Plants Program

University of Wisconsin, Madison
1630 Linden, Madison, WI 53706
(800) 462-7417; fax: (608) 263-0744
www.fastplants.org.

The Wisconsin Fast Plant™ kits allow students to perform seed tests that include reproduction or yield. Information about Fast Plants and instructional materials using them are available under the Fast Plants Learning Ecosystem Web site. ■

## Agencies and Organizations

### American Horticultural Society

7931 East Boulevard, Alexandria, VA 22308
(703) 768-5700, www.ahs.org

Since 1922, the American Horticultural Society has been working to help people become successful and environmentally responsible gardeners. It offers a variety of educational programs for all ages, including the National Youth Garden symposium. At this annual event, teachers and educators learn how to create gardens for young people and how to incorporate those gardens into school curriculum.

### Botanical Society of America

1735 Neil, Columbus, OH 43210
(614) 292-3519, www.botany.org

All types of plant biologists are part of this umbrella organization, which aims to use knowledge of plants, algae, and protists to help solve some of the world's problems. Founded by American botanist Nathaniel Lord Britton, the society publishes a variety of periodicals dedicated to botany, including the *American Journal of Botany.* ■

**www.discoveryschool.com**

The URLs listed here can be accessed through www.discoveryschool.com for your convenience. Go to **Science Collections** and follow the links.

# Questions and Answers

*Common questions about Plants from curious kids*

Discovery Channel School

## Glad You Asked

**Students will love knowing the answers.**

### What is the first plant that ever existed?

According to scientists, land plants have existed on Earth for about 395 million years. *Cooksonia* may be the first land plant. A fern ally with a 1-inch-wide upright stem that forked with two spore-bearing cases at the tips, it lacked leaves, flowers, and roots and was a vascular plant. *Cooksonia* is now extinct and can only be found in fossilized form.

### How did leaves evolve?

Leaves evolved from stems, especially branches. Over time numerous bunches of branches fused together to make a flat surface, perfect for capturing sunlight. This adaptation gave plants an advantage for making and storing food.

### How do plants become coal?

About 300 million years ago, huge tree ferns began to grow on Earth. Forests grew in swamps in what is now North America. When these ferns died, they sank to the bottom of the swamps. For millions of years, dead tree ferns and other plants decomposed there. When the swamps dried up, the layers of dead plants dried up, too, forming many compact layers. Eventually these layers turned into coal. This process continues today.

### Why do certain plants smell stronger at night?

Brightly colored petals, sweet nectar, and fragrance are present in plants for one reason: to attract insect pollinators. Plants such as the honeysuckle depend on moths to pollinate, so they emit a strong, sweet smell at night to lure those nocturnal insects. ■

Name ______________________ Class ______________________

**Diagram of a Dicot**

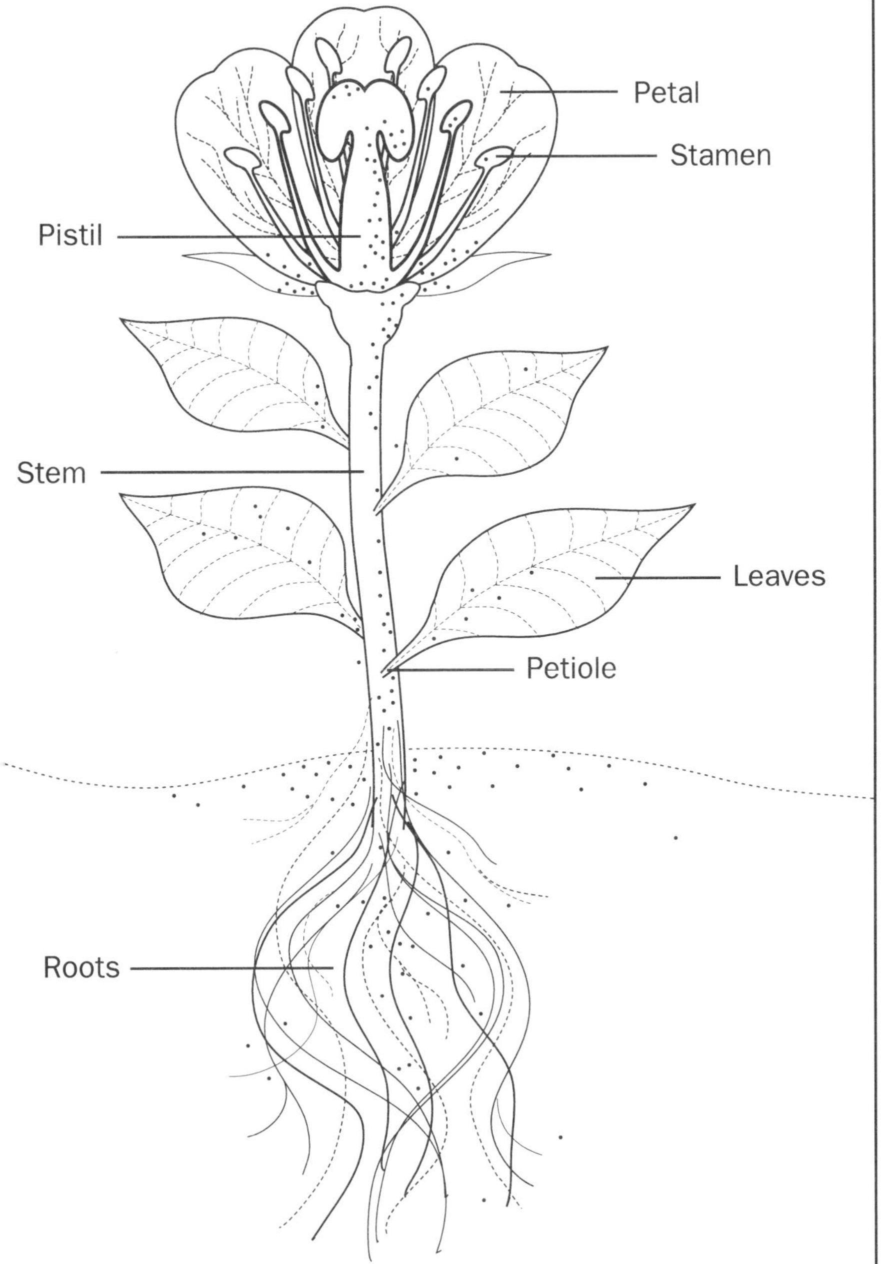


## Monocots & Dicots

Flowering plants, or angiosperms, make up the largest number of plants. About 250,000 species are flowering plants; they show incredible diversity. Flowering plants can be divided into two groups, monocotyledons, or monocots, and dicotyledons, or dicots. The names of the classes refer to the number of a plant's cotyledons, or seed leaves. Monocotyledons have one seed leaf; dicotyledons have two.

Grasses and other monocots usually have long, narrow leaves. Only about 40,000 flowering plants species are monocots. Their flowers grow in threes, or multiples of threes. Besides grasses, monocots include bananas, palm trees, and bulb plants such as tulips and orchids. Most flowering plants are dicots, which usually have broad leaves. Maple and other dicot leaves have networks of veins, many branching off from a central vein. Their flowers usually have parts in sets of fours or fives.

1. Identify the primary physical differences between monocots and dicots.
2. Using the information provided, give three examples of common monocot species and explain your answers.
3. Using the information provided, give three examples of common dicot species and explain your answers.
4. Monocots are believed to have developed from dicots between 80 and 120 million years ago. Some scientists believe monocots started out as underwater species with broad leaves. The leaves of these early monocots eventually became just a base and a petiole, or the part of the leaf that connects to the stem. It transports water to the leaf. Later the leaves developed into their present grasslike shape. Why do you think an underwater environment would cause the leaves to change shape, from a broad leaf to just a petiole? ■

Name ______________________ Class ______________________

## Leaves and Cones

Deciduous trees shed their leaves in the fall to conserve water and food during the winter months. The leaves die because they are sealed off from the tree, the source of its nutrients. Beech, sycamore, and birch trees are deciduous.

Coniferous trees have needles and woody cones, where their seeds are produced. Most coniferous trees are evergreens, so they do not lose their leaves. The cones keep the seeds safe, sealed inside the dry, warm cone until the weather is dry, at which time the cone will open and close its scales. A conifer's needles are designed to help the trees survive cold weather. They are waxy to retain water and flexible to bend under heavy snowfall.

**Sugar maple**

**Spruce**

1. Why might some deciduous trees have trouble surviving in very cold regions?
2. Many animal species exhibit winter-storage behavior that is similar to the behavior of trees. Name two animals that survive the winter in the same way as deciduous trees. Explain your choices.
3. Name two animals that survive the winter in the same way as coniferous trees. Explain your choices.
4. Conifer seeds are kept safe in cones until the conditions are right for dispersal. How does this method of seed dispersal differ from that of deciduous trees? ■

# Science Projects

*Encourage students to discover their own answers*

**National Science Education Standards**

- Students in grades 5–8 should be helped to acquire scientific knowledge, use high-level reasoning, apply their existing understanding of scientific ideas, and communicate scientific information.

## Science Project Guidelines

Science projects should provide students with an opportunity to apply what they have learned about the scientific process and inquiry. Parents and teachers should serve as resources, providing guidance and reaction, but students must own the project, making their own decisions regarding the question, data collection, variables they will control, depth of study, and the format of the final report or display.

When designing their project or investigation, students should:

- Use appropriate tools and techniques to gather, analyze, and interpret data.
- Develop descriptions, explanations, predictions, and models using evidence.
- Think critically and logically to identify the relationships between evidence and explanations.
- Recognize and analyze alternative explanations and predictions.
- Communicate procedures and explanations.
- Use mathematics.

## Science Project Starter Questions

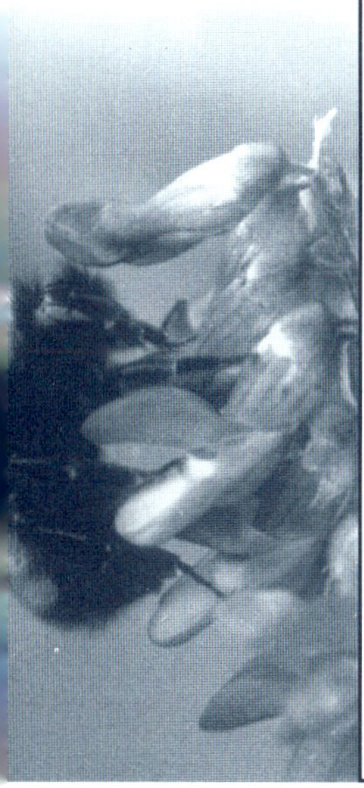

For many students, the hardest step in a science project is selecting the topic and posing the question. The questions should be related to content previously studied, in or out of school, and be the result of genuine curiosity. Projects do not have to be invasive; they do not have to test or experiment with pollution, result in injuring or killing organisms, or impact environment. These questions can be used to provide ideas for investigations.

**Research**

Which plants are indigenous to my region? Select one native species, and answer the following:

- How does it become pollinated?
- What is its root structure?
- How is it used by animals? by humans?
- What is its life cycle?

**Modeling**

How do plants that do not have flowers reproduce?

- What makes the bee a good pollinator?
- How do grasses such as corn and wheat produce flowers and reproduce?

**Experiments**

What are plant tropisms? How do plants react to different stimuli?

- How do different pollinators affect seed production?
- How do plants respond to different amounts of carbon dioxide in the air? ■

## Young Scientist Challenge 2000

Discovery Channel School provides assistance to teachers and students at http://school.discovery.com/sciencefaircentral with an online science fair handbook.

Remember that the objective of a science project is to learn more about science, not to produce an elaborate display or win a prize.

For students who want the ultimate challenge, enter one of these projects or any others in the Discovery Young Scientist Challenge. Visit www.discoveryschool.com/dysc to find out more. ■

# T Testing

*How to assess student performance and concept mastery*

Discovery Channel School

**Performance-based evaluation enables the teacher to assess student understanding of concepts through the use of activities, experiments, projects, and journal writing. Performance assessment consists of a task and rubrics—a scoring method used to judge the quality of student performance.**

## Student Journals

**Activities** and **Experiments** in *Plants A-Z* can be used as both teaching and testing experiences. Student journals enable you to use these experiences to obtain both pre- and post-evaluation of student learning. Prior to each experience, pose a question and ask students to write their thoughts about that question. During the experience, have them record notes and data. Following the experience, students reflect on what they did and what they now know about the question asked. This pre- and post-evaluation will provide a documentation of student growth in understanding the big ideas about plants.

## Developing a Scoring Rubric

Because student output, such as a multimedia slide show, is more complex than a written assignment, assessing a student's work may seem more difficult. In a project such as this, students must be able to express their learning clearly, demonstrating not just what they know but also their ability to communicate it to others. An important part of assessing the work consists of students evaluating themselves and each other in addition to evaluation by the instructor. The tool that makes it possible to evaluate work in this student-centered context is a rubric.

A rubric is simply a list of criteria by which student output is judged. It defines the attributes of a quality project, provides the criteria by which a project will be assessed, and gives specific indicators that describe various levels of achievement for each of those criteria. In this respect, a rubric is a concrete guide or checklist that students can refer to as they work, besides being a measuring device for the teacher.

For example, a generic rubric for a complex project might look something like this:

| | Needs Work (1 point) | Average (2 points) | Best (3 points) |
|---|---|---|---|
| **Content** | Data is inaccurate in places or key pieces of information are missing. | Data is complete, but inaccurate in places. | Data is both accurate and comprehensive. |
| **Presentation** | Is drab and ordinary. | Is entertaining and interesting. | Shows creativity and is entertaining; shows that the student had fun putting it together. |
| **Effort** | Project exhibits minimal effort to fulfill the assignment. | Project shows personal interest in the topic and thoughtful analysis. | Project shows evidence of personal interest, thoughtful analysis, and creativity in the information's presentation. |
| **Publishing** | Text has numerous errors in grammar, usage, spelling, or punctuation. | Text has few errors in grammar, usage, spelling, or punctuation. | Text has no errors in grammar, usage, spelling, or punctuation and utilizes higher level vocabulary and complex sentence structure. |

This example is just a suggestion. Keep in mind that you know your students best and you know their academic needs. As the teacher, you will need to develop your own rubric for your class so that it reflects:

- the age and prerequisite skills of your students
- the specific goals you have for your class
- the complexity of the content topic selected
- the level of achievement you expect

The science project questions on page 23 can serve as long-term tasks or the basis for multimedia or similarly complex projects.

## Plants CD-ROM

Students can explore the integral role plants play in their lives through this CD-ROM by Discovery Channel School (#743625). To order, call (888) 892-3484. ■

**Will your students be the ones to find the answers someday?**

## Plants as a Cure for Disease

A willow tree can get rid of your headache. Not directly, but it may help indirectly. Aspirin, a common headache remedy, was originally made from chemicals extracted from willow trees. Today aspirin is manufactured in a laboratory from chemicals derived from those in the willow tree. Other common lifesaving drugs are still made from plants. Like aspirin, many others are created synthetically. Yet to be discovered are many more plant species.

Until recent history, many medicines came directly from plants. And in many parts of the world, this is still the case. The first pure plant substance was morphine. About 100 years ago, morphine was extracted from the opium poppy, a flowering plant. Morphine has been invaluable in providing pain relief for people ever since. Digitalis, widely used to treat heart disease, comes from foxglove, a poisonous plant. Quinine is extracted from the cinchona tree and is used to treat malaria. The Madagascar rosy periwinkle is beneficial in the treatment of childhood leukemia, a form of cancer.

Today scientists are rediscovering these and many other plants in the rain forest, where every day thousands of trees are cut down. Many valuable plant species may never be discovered.

Scientists are not certain of the cures plants hold. Using plants for healing is tricky. Many healing plants are deadly in the wrong dose. Scientists must work carefully, but quickly, to test new medicines. The rain forests and other wilderness areas may carry a plant that cures AIDS, cancer, or many other diseases. We may not know for a long time.

## Repairing the Ozone Layer

On the news we often hear dire warnings about the vanishing ozone layer. Not many people could explain what the ozone layer is and how it's being damaged, but many people could probably tell you that plants and trees are good for it.

Ozone is a gas, created by the reaction between the Sun and oxygen. Made of oxygen, the ozone layer is a section of the atmosphere. It is crucial to the survival of living things because it filters the Sun's cancer-causing radiation. Most oxygen is a by-product of photosynthesis.

In the 1970s, scientists discovered that chlorofluorocarbons (CFCs)—chemicals released into the atmosphere by refrigerators and aerosol sprays—were damaging the ozone layer. When CFCs reach the ozone layer, they react with sunlight and release chlorine, a chemical that destroys ozone molecules.

In 1987, 36 countries signed the Montreal Protocol, a treaty for the protection of the ozone layer. Since then many countries have taken additional steps to stop producing CFCs and other ozone-depleting chemicals. However, even if all production of these chemicals stops, those already in the atmosphere will continue to damage the ozone layer for many years. Scientists hope the ozone layer will mend itself in the future, but they don't know when or if that will happen. ■

# V Vocabulary

*Key Plants words and terms*

**Use for your own reference or copy and distribute to students.**

**chlorophyll**—a green substance found in plant cells that is necessary for the process of photosynthesis

**chloroplast**—tiny green structures inside plant cells where photosynthesis takes place

**conifer**—tree that bears woody cones

**deciduous**—plant that loses it leaves in the winter

**dicotyledons**—flowering plants with two seed leaves

**epidermis**—an outer layer of cells

**fruit**—structure that contains seeds developing from a pollinated flower

**margin**—border of a leaf

**monocotyledons**—flowering plants with one seed leaf

**petiole**—part of a plant that connects the leaf to the stem

**photosynthesis**—process by which green plants make their own food

**pistil**—female part of a flower, consisting of the stigma, style, and ovary

**pollination**—process by which pollen is transferred from the male part of a flower to the female part of a flower

**stamen**—male part of a flower where pollen is produced

**stigma**—female part of a flower where pollination takes place

**stomata**—tiny holes in the epidermis of a leaf through which gases travel

**veins**—structures that strengthen plants and carry nutrients to different parts of the leaf

**xylem**—system that transports water and nutrients from the roots to other parts of the plant ■

# Writing Ideas

*Reproduce and distribute for instant topic-related writing assignments*

**Assess students' understanding of plants through their writing on the topic.**

## Here on Earth

Green plants are responsible for the survival of all animals, including humans. Write a letter to send on the space shuttle. Explain to a being from another planet the role plants play in our planet's survival. Be sure to explain the different ways plants make it possible for animals (including humans) to live on Earth. ■

## Your Own Walden Pond

Nature inspired great American writers like Emily Dickinson, Henry David Thoreau, and Robert Frost. Take a nature walk, taking time to observe the plants you see. Bring a notebook to record your observations. When you finish, find a quiet place to work and let nature inspire your writing. You may want to compose a poem or set a short story in the outdoors. Or you may want to simply describe your observations. ■

## The Final Frontier or the Last Hiding Place?

Scientists have identified more than two million living species in the world. That may seem like a lot, but some scientists believe that as many as eight million more may exist. Many of these unknown species live in areas of the earth that have not been explored or have been explored very little, such as the deepest parts of the oceans and the depths of the Amazon rain forests. Some people believe that we should leave these areas alone. Others believe that scientists should explore every last place on Earth. What do you think? Research this topic and take a position. Then write a paper reporting on your view. Set up a debate with a classmate who has another opinion. ■

# X Marks the Spot

*A geography link opens up worlds of discovery*



## Forests and Deforestation Around the World

Forests once covered most of the Earth. Today forests have been lost to farmland, housing, and the logging industry. Many organizations, such as the Sierra Club, the Nature Conservancy, and Global Forest Watch, are devoted to saving forested lands. Despite their efforts, deforestation continues every day at a rapid rate. The United States protects forested land through government and private efforts, yet it is a continual struggle. Some countries don't have many resources, and their governments may have to choose between saving the forests or protecting the economy. Conservation groups around the world hope to find a way to encourage poorer countries to save forested land without sacrificing the health of their economy.

The map below shows the countries that have the most forested land, as well as the countries that have cut down the most forests since 1981. ■

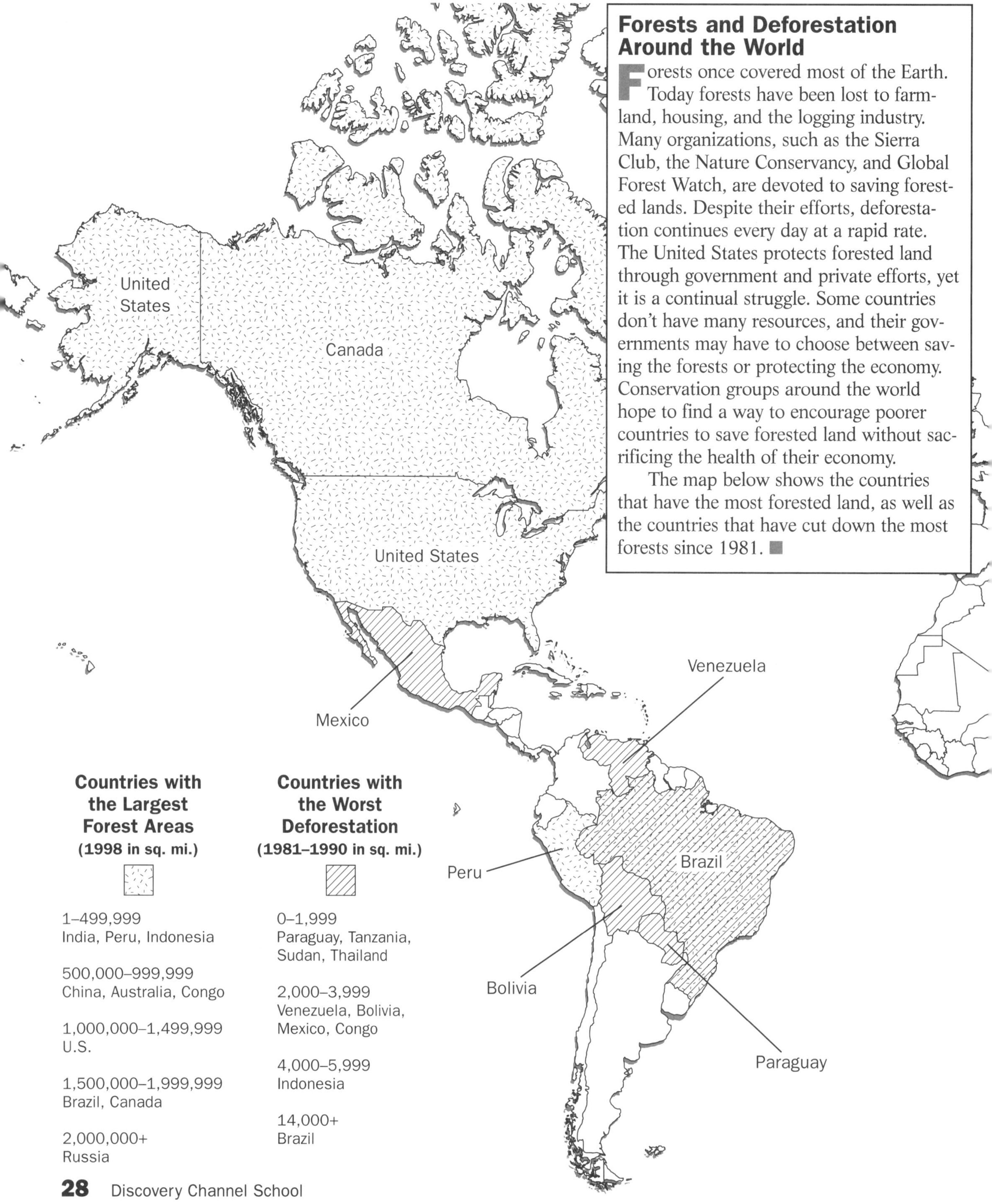


**Countries with the Largest Forest Areas (1998 in sq. mi.)**

1–499,999
India, Peru, Indonesia

500,000–999,999
China, Australia, Congo

1,000,000–1,499,999
U.S.

1,500,000–1,999,999
Brazil, Canada

2,000,000+
Russia

**Countries with the Worst Deforestation (1981–1990 in sq. mi.)**

0–1,999
Paraguay, Tanzania, Sudan, Thailand

2,000–3,999
Venezuela, Bolivia, Mexico, Congo

4,000–5,999
Indonesia

14,000+
Brazil

# X Marks the Spot

**For students:**

1. How many countries with the largest areas of forest are also among the top deforesting countries? What could you infer from your answer about the majority of the countries with large areas of forest?
2. According to the information on the map, what country is doing the best job protecting forests? Explain your answer.
3. Which countries have some of the most forested areas but are also among the worst deforesters?
4. How could the United States work with its neighbor Mexico to promote forest conservation?
5. What are some reasons you think South America does not do as well as North America as a whole in terms of protection of forested land?
6. Where in the United States do you think the largest areas of forest are? Why? ■

# Y Year After Year

*Significant science, traced over time*

Discovery Channel School

Plants have existed for millions of years, yet the study of plants has only been undertaken for a few thousand years. From ancient ferns to healing herbs to today's genetically engineered food crops, scientists continue the quest for knowledge about the plant world. The following is a chronology of the history of plant study. ■

**About 400 million years ago**—Plants begin to grow on land.

**About 300 million years ago**—Forests of giant tree ferns begin to grow on land.

**About 225 million years ago**—The first conifers (spruce, fir, and pine trees) thrive.

**About 125 million years ago**—Flowering plants evolve.

**About 300 BC**—Greek philosopher Theophrastus studies and writes about plants, describing almost 500 species. He distinguishes flowering from nonflowering plants and recognizes that plants have male and female parts.

**About 77 AD**—Roman natural historian Pliny the Elder writes *Historia Naturalis* (Natural History). This work was the authoritative source of information about plants and many other topics for 1,000 years.

**About 77 AD**—Greek physician Dioscorides writes *De Materia Medica* (The Medical Material), in which he describes over 600 plant species and their medicinal uses.

**1650–1651**—Johann Bauhin's *Historia Platarum Universalis* (Universal History of Plants) is published. Bauhin's work contained detailed descriptions of more than 5,000 plants. Bauhin was the first to develop the concept of genus. He was Gaspard Bauhin's brother. (See Carolus Linnaeus biography, page 17.)

**1753**—Carolus Linnaeus's *Species Plantarum* (1753) is published. Earlier, *Systema Naturae* (1735) was published. In these, he creates a system of classification for all living things.

**1809**—Jean-Baptiste Lamarck publishes *Philosophie Zoologique* (Zoological Philosophy). He explains his ideas about the effect of the environment on plants and other living things.

**1859**—Charles Darwin and Alfred Wallace set forth the theory of natural selection.

**1865**—Gregor Mendel publishes *Experiments with Plant Hybrids,* in which he describes how traits are inherited. His work would not be accepted and appreciated until 35 years later.

# Year After Year

**1876**—Darwin is the first to give a complete analysis of the concept of "hybrid vigor" in his book *Cross and Self-Fertilization with the Vegetable Kingdom.*

**1915**—*Die Natürliche Pflanzenfamilien* (The Natural Plant Families) is published. Its authors, A. Engler and K. Prantl, present a classification of all plants. Though the classification has some basic errors, its breadth makes it an important resource even today.

**1956**—Gregory Goodwin Pincus and M.C. Chiang discover that dioscorea, a drug made from the yam, prevents conception, and the birth-control pill is developed.

**1972**—DDT usage as a pesticide is banned in the United States.

**1973**—The Endangered Species Act is passed in the United States to protect plants and animals.

**1983**—The first "genetically modified" plant is created. Made by transferring a specific gene from one plant to another, the purpose is to create a plant with specific characteristics.

**1990**—NASA and the George Peck Seed Co. start Project SEEDS. Students compare seeds exposed to the conditions in outer space to seeds stored on Earth.

**1993**—The U.S. Food and Drug Administration announces the approval of the growth hormone bovine somatotropin, a drug used to increase milk production in cows. It proves highly controversial and receives extensive media coverage.

**1999**—A team of researchers present genetic evidence that *Amborella* may be the most primitive living flowering plant.

**2000**—The United States Congress takes another look at the Endangered Species Act to determine whether it's effective. ■

# Z Zingers

*Thought-provoking tidbits*

Discovery Channel School

## Did You Know?

**A great way to begin or end class.**

- More than 400,000 known plant species exist today. That may seem like a lot, until you compare that number to the number of animal species—more than one million.
- The largest flower in the world stinks. Found only in Southeast Asia, *Rafflesia arnoldii* can grow up to 3 feet in diameter. A parasitic plant (no roots or stems), it grows attached to a vine or another host plant. It smells like rotting meat to attract flies, the insects that spread its pollen.
- Do you know when you're eating a plant? Not necessarily. When you eat candy, you are also eating a plant. Candy is mostly sugar, which comes from the sugarcane, a plant that grows in tropical areas.
- People eat different parts of plants: flowers, leaves, seeds, fruit, and roots. Broccoli and cauliflower are flowers. Lettuce and spinach are leaves. Peanuts without the shells and peas without the pods are seeds. Apples, tomatoes, and eggplants are fruits. Carrots and radishes are roots.
- The leaves of many trees are bright orange, red, and yellow all year long, not just in the fall. We just can't see the bright colors until the tree gets ready for winter and stops sending food to its leaves. Without food, the leaves lose their green color, and the bright colors that were there all along finally show themselves. Later the leaves die and fall from the tree.
- The spines on a cactus do more than protect it from predators. In its dry desert environment, a cactus uses spines for survival. They collect water and moisture, which the cactus absorbs.
- About 400 plants are carnivorous. These plants do not eat meat, but they do trap animals—mostly insects. The best known is probably the Venus flytrap. As its name implies, the plant traps flies and other insects and ingests them for their nitrogen and potassium. The plant attracts insects with its honey-coated leaves. Each trap is a single leaf. When the trigger hairs on a Venus flytrap's traps detect an insect, the traps snap shut. The plant produces enzymes that help it digest its prey.
- Throughout history, people have used plants for purposes that today seem strange. Women in ancient Greece rubbed their cheeks with mullein plant as a sort of makeup: The plant's fuzz left an irritant on the women's cheeks, causing them to glow.
- Some poisonous plants have tasty but harmless relatives. The plants that bear mangoes and cashew nuts are relatives of poison ivy. ■

**Credits:**
**Suzy Wall Sensbach, Joyce Lowry Weiskopf, Ph.D.,** Writers
**Katie King,** Editor
**Illustrations:**
p. 10, photosynthesis/cross section of a leaf; reproductive parts of a flower, David F. Kelly; p. 22, sycamore leaf; conifer; pine tree outline; forest map and key, David F. Kelly; map, pp. 28–29, Thomas Gagliano
**Cover:** PictureQuest

### Answers

**All About Plants Crossword, p. 9**

| Across | Down |
|---|---|
| **1.** chlorophyll | **1.** lichen |
| **2.** coniferous | **2.** root |
| **3.** Venus flytrap | **3.** stamens |
| **4.** algae | **4.** stigma |
| **5.** deciduous | **5.** epidermis |
| **6.** photosynthesis | **6.** xylem |
| **7.** dioxide | |
| **8.** leaf | |

**Poison Plants Mix-up, p. 9**

| | |
|---|---|
| **1.** azalea | **4.** mistletoe |
| **2.** daffodil | **5.** holly |
| **3.** foxglove | **6.** poison ivy |